YOUR KNOWLEDGE HAS VALUE

- We will publish your bachelor's and master's thesis, essays and papers

- Your own eBook and book - sold worldwide in all relevant shops

- Earn money with each sale

Upload your text at www.GRIN.com and publish for free

Elguja Medzmariashvili, Leri Datashvili, Nikoloz Medzmariashvili, Otar Tusishvili, Ariadna Jakhua, Mamuka Sanikidze

V-fold bar deploybale ring with deployable bearing ring

GRIN Verlag

Bibliografische Information der Deutschen Nationalbibliothek:

Die Deutsche Bibliothek verzeichnet diese Publikation in der Deutschen Nationalbibliografie; detaillierte bibliografische Daten sind im Internet über http://dnb.d-nb.de/ abrufbar.

Dieses Werk sowie alle darin enthaltenen einzelnen Beiträge und Abbildungen sind urheberrechtlich geschützt. Jede Verwertung, die nicht ausdrücklich vom Urheberrechtsschutz zugelassen ist, bedarf der vorherigen Zustimmung des Verlages. Das gilt insbesondere für Vervielfältigungen, Bearbeitungen, Übersetzungen, Mikroverfilmungen, Auswertungen durch Datenbanken und für die Einspeicherung und Verarbeitung in elektronische Systeme. Alle Rechte, auch die des auszugsweisen Nachdrucks, der fotomechanischen Wiedergabe (einschließlich Mikrokopie) sowie der Auswertung durch Datenbanken oder ähnliche Einrichtungen, vorbehalten.

Imprint:

Copyright © 2013 GRIN Verlag GmbH
Druck und Bindung: Books on Demand GmbH, Norderstedt Germany
ISBN: 978-3-656-48589-6

This book at GRIN:

http://www.grin.com/en/e-book/231704/v-fold-bar-deploybale-ring-with-deployable-bearing-ring

GRIN - Your knowledge has value

Der GRIN Verlag publiziert seit 1998 wissenschaftliche Arbeiten von Studenten, Hochschullehrern und anderen Akademikern als eBook und gedrucktes Buch. Die Verlagswebsite www.grin.com ist die ideale Plattform zur Veröffentlichung von Hausarbeiten, Abschlussarbeiten, wissenschaftlichen Aufsätzen, Dissertationen und Fachbüchern.

Visit us on the internet:

http://www.grin.com/

http://www.facebook.com/grincom

http://www.twitter.com/grin_com

V-FOLD BAR DEPLOYABLE RING WITH DEPLOYABLE BEARING RING

Authors: Leri Datashvili[1]; Nikoloz Medzmariashvili[2]; Otar Tusishvili[2]; Ariadna Jakhua[2], Mamuka Sanikidze[2]

[1] Institute of Lightweight Structures, Munich Technical University, D-85747 Garching, Germany
[2] Institute of Constructions, Special Systems and Engineering Maintenance of Georgian Technical University. Address 68b Kostava st. 0175. Tbilisi. Georgia.

Keywords: V-fold bars, ring, deployable, ring,

Abstract: The mentioned works were offered by European Space Agency, whose employees are the inventors of the mentioned system. The real structure was performed in this work. The deployment of V-fold bar that itself causes deployment of deployable ring is reached by use of constant torque springs. The deployment stabilizing mechanism is located in joints. Deployable bearing ring that also tensions pre-stressed spatial rod system can be as conical also prism.

Intoduction: Georgian Technical University and Munich technical University in accordance of ESA patent are jointly working on practical realization of new schemes of the large deployable space reflectors [1][2].

The deployment of reflector is carried out by application of deployable bearing ring. Should be mentioned that the use of deployable bearing ring in large deployable reflectors counts more than 20 years [3][4].

The large scale test of large deployable reflector created in accordance with the mentioned system was performed on orbital station "MIR" [5].

The following researches are performed to create new schemes of large deployable reflectors with deployable rings that are generally directed on its lightening [6][7][8][9].

In order to improve the deployable reflector was created conical reflector with V-fold bar deployable ring [10][11][12].

The body of the article

The general views of the reflector antenna having "V-fold bars" in deployed and transportable states can be seen in Figure 1.

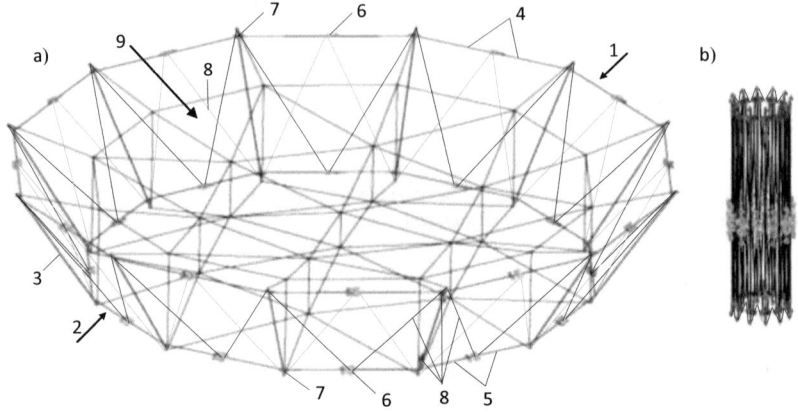

Figure 1. General view of the construction

a) in the deployed state and b) in folded state.
1. Upper ring; 2. Lower ring; 3 Post; 4. V-fold bars of the upper ring; 5. V-fold bars of the lower ring; 6. Hinge that connects the V-fold bars, provided with a permanent force spring mechanism; 7. Synchronization mechanism that connects hingedly the sections; 8. Deployment stabilization system; 9. Central part.

The conical transformable load-bearing ring itself which is provided with "V-fold bars" comprises the following:

- Upper and lower rings connected by posts;
- Hinges for connecting the V-fold bars with permanent moment spring deployment mechanisms;
- Hinges for connecting the sections with synchronization mechanisms;
- Deployment stabilization system.
    The central part comprises the following:
    - Functional and technological elastic meshes;
    - Vertical or inclined tensioners for connecting them.

A kinematical diagram of deployment/folding of a single section of the space reflector antenna load-bearing ring and its geometrical parameters are shown in Figure 2.

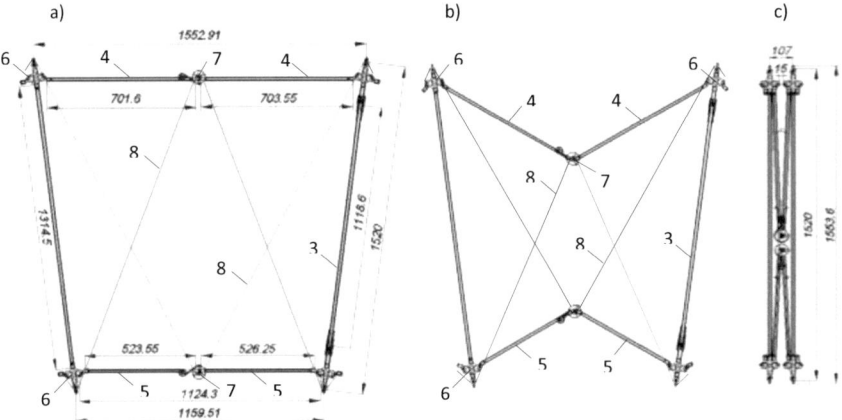

Figure 2. Kinematical diagram of deployment/folding of the load-bearing ring single section.
a) Deployed state; b) Partially folded state; c) Folded state; 3 Post; 4. V-fold bars of the upper ring; 5. V-fold bars of the lower ring; 6 Synchronization mechanism; 7. Hinge with permanent force spring deployment mechanisms; 8. Deployment stabilization system.

The deployment of the load-bearing ring is effectuated by permanent moment flat springs.

Due to the symmetrical distribution of forces, coupled permanent moment spring mechanisms are used on each pair of V-fold bars (Figure 3).

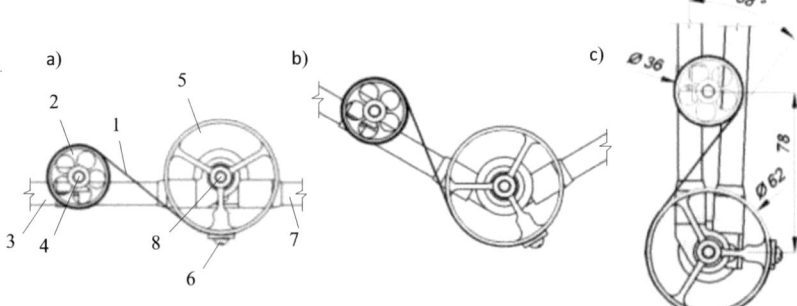

Figure 3. Permanent moment spring deployment mechanism.

a) Deployed state; b) Partially folded state; c) Folded state; 1. Spring; 2. Small drum; 3. Left V-fold bar; 4. Rotary axis; 5. Coiling drum; 6. Bolt; 7. Right V-fold bar; 8. Motionless axis.

The horizontal rods are interconnected by means of synchronous transmission mechanisms (Figure 4), which allows simultaneous motion of the levers contained in the upper and lower rings.

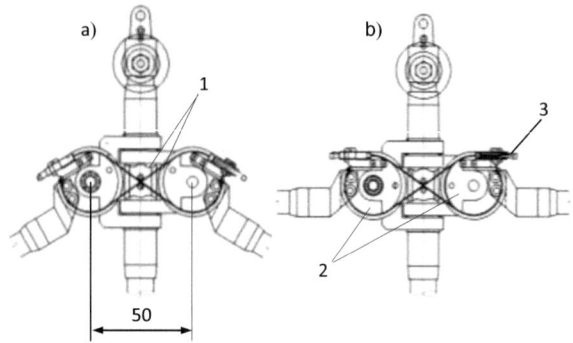

Figure 4. Synchronization mechanism.

a) – Partially folded state; b) – Deployed state; 1. Cable; 2. Guide roller; 3. Cable stretching device.

Together with the advantages, deploying ring having "V-fold bars" and a spring is characterized by some disadvantages.

1. The springs that determine the ring deployment energetics, have considerable weight.

2. The deployment stabilization system which is formed by continuous cables on several sections makes the ring deployment more difficult due to asymmetrical forces existing in the cables passing over the levers. Moreover, it needs both electric drive and a system of rollers disposed in the units and near the units, which makes the construction even heavier and more complex.

3. The deployment stabilization system having a cable does not create the geometrical unchangebility condition in the deployed ring because the sufficient stretching force cannot be produced and a symmetrical stressed picture is not created therein in spite of its geometrical disposition.

4. The complex stressed and deformed picture at the ring deployment stages, including at the final stage, does not exclude "straightening of all V-fold bars". Very often a single "V-fold bar in folded state" is remained. Therefore it is just the bar which ensures the force that is sufficient for deploying the ring as well as for pre-stressing the elastic central part as a whole. It is possible to eliminate this effect but this will require additional structural modifications to be accomplished in the ring construction.

It should be noted that the weight of the reflector antenna having "V-fold bars" wherein permanent moment springs are used as a deployment mechanism is 16kg.

In spite of designing several different modifications, the deploying ring still needs to be optimized, at least by introducing some modifications in its line diagram, due to the partial drawbacks thereof.

To this end, the reflector antenna having "V-fold bars" that has been proposed by the European Space Agency has been designed and tested on deployment not only by use of springs but also by electric drives and a deploying cable. These possibilities were already envisioned in the reflector antenna having "V-fold bars"(see Figure 2 and Figure 4).

References:

1. E. Medzmariashvili, Sh. Tserodze, O. Tusishvili, N. Tsignadze, J. Santiago-Prowald, C.G.M. Van't Klooster, N. Medzmariashvili. Mechanical Supporting Ring Structures CEAS Space Journal of European Aerospace Sicieties. ISSN 1868-2502. Published online. June 2013
2. E. Medzmariashvili, N Tsignadze, Sh. Tserodze, J. Santiago-Prowald; C. Mangenot, C.G.M. Van't Klooster, H. Baier, M. Janikashvili. Design of Reflector with Double Pantograph and Flexible Center. Proceedings of ESA Antenna Workshop on Large Deployable Antennas. 2-3 October 2012. ESTEC, Noordwijk, The Netherlands.
3. E. Medzmariashvili, A. Iacobashvili, G. Bedukadze. Creating and Testing of Large Space Structures of High Precision Surface. Space Power, Volume 12, Number 1-2, 1993.
4. E. Medzmariashvili. Transformable Space and Ground Structures. Monograph. Pub. Georgia-Germany-Liechtenstein. 1995
5. E. Medzmariashvili, V. Blagov, A. Chernyavsky. A Space Eperiment Confirms Reflector's High Reliability. Aerospace Courier, No 6, 1999
6. E. Medzmariashvili. Deployable Space Reflector Antenna. "E.V.M". International Publication No WO03/003517 A I. 9.01.2003. International Application Published Under the Patent Cooperation Treaty (P.C.T.)
7. E. medzmariashvili, Sh. Tserodze, V. Gogilashvili. New Variant of the Large Deployable Ring-Shaped Space Antenna. Space Commmunications 22 (2009) 41-48.
8. E. MEdzmariashvili. The Basic Principles of the Large Deployable Space Antenna. Transactions Technical University of Georgia. 2 (472) 2009
9. E. Medzmariashvili, N. Medzmariashvili. Constructive Logic of Reflector Created with Double Pantograph Deployable Load-Bearing Ring. Proceedings of ESA Antenna Workshop on Large Deployable Antennas. 2-3 October 2012. ESTEC, Noordwijk, The Netherlands.
10. E. Medzmariashvili, N. Medzmariashvili, O Tusishvili, N. Tsignadze, J. Santiago-Prowalds, C. Magenot, H. Baier, L. Scialino, L. Philipenko. The possible Options of Conical V-fold Bar Ring's Deployment with Flexible Pre-Stressed Center. CEAS Space Journal of European Aerospace Sicieties. ISSN 1868-2502. Published online. June 2013
11. E. Medzmariashvili, L. Datashvili, J. Santiago-Prowald, L. Scialino, H. Baier, C. Mangenot, O. Tusishvili, N Tsignadze, K. Chikvaidze, M. Janikashvili. The Structure of Conical Reflector with V-fold Bar's Deployable Ring. Proceedings of ESA Antenna Workshop on Large Deployable Antennas. 2-3 October 2012. ESTEC, Noordwijk, The Netherlands.

12. E. Medzmariashvili, J. Santiago-Prowald, C. MAngenot, H. Baier, L. Scialino, L. Philipenko, N. Medzmariashvili. The Possible Options of Conical V-fold Bar Ring's Deployment with Flexible Pre-Stressed Center. Proceedings of ESA Antenna Workshop on Large Deployable Antennas. 2-3 October 2012. ESTEC, Noordwijk, The Netherlands.